The
Art of Sundial
Construction

For Màiri, my Inspiration;
and for Susannah,
my First Serious Pupil.

i

Susannah Rose Andrews

The Author

The Art of Sundial Construction

by

Peter I. Drinkwater

Published by P. Drinkwater, 56 Church Street,
Shipston-on-Stour, Warwickshire,
England.

1985

ISBN 0 946643 09 1

Set by Mr. S. C. Bloomfield

Compositor for Hand Setting Mr. D. J. Harris

Machinist Mr. R. A. Beckett

Printed by Bloomfield & Son, Grove Road,
Stratford upon Avon.

Designed by Peter I. Drinkwater.

iv

ACKNOWLEDGEMENTS

Are due to:

Miss Màiri R. Macdonald; who has carefully read through the Manuscript of this Work, making many helpful Suggestions for its improvement (including the Idea that the Diagrams would be made clearer by the printing of the Reference Letters in Red), all of which have been adopted, and has inspired the production of the Diagram on Page: 4.

Miss Margaret Shepard; who has thoroughly revized the Punctuation of the Work throughout its Length, has pointed out many Grammatical Errors and Spelling Mistakes therein, and has also (to the Author's great pleasure) substantiated the Design on Page: 37 as a splendid Embroidery.

Miss Susannah Rose Andrews; for whose Benefit and Education the Diagrams on Pages: 30, 44, and 46 were specifically produced, and who has most kindly contributed further the Drawing of myself which forms the Frontispiece of this little Book.

Mr. David Andrews; at whose specific Request the Diagram on Page: 56 was originally delineated; so that he might himself construct such a Dial.

Mrs. Annemarie Krimke and Mrs. R. J. Bloomfield; whose Work on the Old German Verses has enabled both the Production of the English Translations of the same here included, and the correct realization of the original German Texts which are also incorporated; and to Mr. S. C. Bloomfield for his sensitive handling of the Original Manuscript.

Miss Mary Smith; whose interest in seeing the Publication of this Book has done much to encourage the Author in his work.

Prof. & Mrs. Derrett; who have likewise encouraged the Author; and to Prof. Derrett in particular for pointing out the Feminization of Time in the Verse quoted on Page: 58.

Christopher St. J. H. Daniel, Esq. of Greenwich who specifically requested me to complete and publish this work, as meeting a current need for a Book on such a Subject; and to Mrs. Doreen Bowyer also for her own encouragement in the same vein and to similar effect.

To the Shades of the Anonymous Latin and German Authors of the Verses scattered throughout this Work (from the Sundial Inscriptions quoted in Herman Witekind's *Conformatio Horologiorum Sciotericorum* of 1576 (H.W.), and Sebastian Münster's *Compositio Horologiorum* of 1531 (S.M.)), with apologies for the present Author's "Imitations" in English.

Also to Mrs. Vivienne M. Griffin; who has spent many hours reading through the Proofs of this Work with me.

FOREWORD

SUNDIALS have been constructed since mankind first set about trying to mark out the course of the day. Daylight and darkness provided the basic means by which time was measured, a single period of daylight and darkness being called *a day*. The motion of the sun and the consequent movement of a shadow enabled man to divide up the period of daylight into convenient parts, i.e. *hours,* to suit his various needs. In many instances he would have used his own shadow to do this, but as civilizations developed, he constructed ingenious instruments to serve this purpose. These were probably the earliest of all scientific instruments, and we know them as *sundials*. Ancient Greece and Rome had such sophisticated instruments in everyday use, for determining the passage of time, before the birth of Christ, although they marked what are now known as *seasonal* hours or unequal hours, rather than the uniform system we use today.

In England, as in other European countries, the construction of sundials was brought to a fine art and flourished from the 16th century through to the end of the 18th century. Every scholar was expected to study and understand this mathematical art (or science). We know that such famous persons as Sir Christopher Wren and Sir Isaac Newton, amongst others, were familiar with sundials.

As one might expect with a subject that was once a course of study in its own right, many books have been published about the making of sundials. The earliest English work devoted to the subject was by Thomas Fale, entitled *Horologiographia: The Art of Dialling,* which was published in 1593. It was a popular little book and was reprinted frequently from 1622 to 1652, some sixty years!

Today, sundials still attract people, although few take them seriously or realise their historical importance. They are regarded with interest rather as ornamental antiques than as scientific instruments. But there are some with enquiring minds who succumb to the fascination of the passing shadow or a moving point of light, and who want to know more about sundials, and wish to understand how they are constructed. Whilst there is a wealth of sundial or *dialling* literature from the past, which is available to us through institution libraries, universities and museums, such antiquarian works are hard to come by in everyday life. Unless one is very lucky, it is likely that one would only find an early dialling work through the offices of a specialist antiquarian bookshop. The book would almost certainly be expensive, but, also, it might well prove to be obscure and not all that easy to understand.

It may surprise one to know that a number of modern sundials are produced every year, often unique works of science and art. Likewise, modern sundial literature is published at much the same rate, but nevertheless, it must be said that there are few instructive dialling books in the English language, which are readily available to those who want to learn how to construct a sundial.

Peter Drinkwater might be described as something of an eccentric: he walks where others would drive, he sees what others would miss, he is a man who lives his life close to nature, who loves and understands the past whilst being fully aware of the present. Furthermore, Peter Drinkwater loves sundials, he has studied them and constructed them through the medium of the early diallists, from their various published works. He is an author, artist and publisher, who has produced a number of beautiful books on local social and historical subjects. He has now, I am glad to say, produced this delightful book on the *Art of Sundial Construction*, which not only fulfils a current need for a clear, simple and modern work on the subject, but which also has the charm of those books produced by the great diallists of the past.

Christopher St. J. H. Daniel
National Maritime Museum
Greenwich

28 January 1985

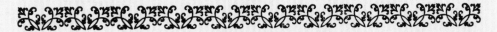

CONTENTS

Venus, the Mother of that bastard Love
  Which doth usurp the World's great Marshal's Name,
Just with the Sun her dainty Feet doth move
  And unto Him doth all her Gestures frame;
  Now after, now afore, the flatt'ring Dame
    With divers cunning Passages doth err,
    Still Him respecting that respects not Her.

For that brave Sun, the Father of the Day,
  Doth love this Earth, the Mother of the Night,
And like a Reveller in rich Array
  Doth dance his Galliard in his Leman's sight,
  Both back and forth and sideways passing light.
    His gallant Grace doth so the Gods amaze
    That all stand still and at his Beauty gaze.

But see the Earth when He approacheth near,
  How She for Joy doth spring and sweetly smile;
But see again her sad and heavy Cheer
  When changing Places he retires awhile.
  But those black Clouds he shortly will exile,
    And make them all before his Presence fly
    As Mists consum'd before his cheerful Eye.

From *Orchestra*, by Sir John Davies (1569-1626).

AUTHOR'S PREFACE

DIALLING is not a suitable Subject for learning by rote: at some Stage, hopefully as early as possible, there must be a Dawning of Understanding as to what is involved, and what it is that is required to be done.

Geometry is not well taught in Schools. What little is taught is either so inconsequential as to be a waste of time; or it is treated as a mere jumping-off point for Mathematical Abstractions which have no practical application. Thus, should anyone nowadays wish to make a Sundial; they are often obliged to start from Second Principles: the Groundwork, the real Basis of the Art, being totally unrepresented in any Standard Textbook. The Present Author was not taught any Geometry at School beyond the Theorem of Pythagoras; and his own early tentative Inquiries into the Art of Dialling were met with such a degree of Incomprehension as to convince him that it was so difficult of attainment as to be completely beyond his powers. Recognizing in this, however, yet another manifestation of the modern tendency to exhort the Pupil to "Behold and Marvel" rather than to "Understand and Emulate", he has proceeded privately with his Studies, aided more by the dusty Writings of the long dead than by any putative assistance or encouragement from the living, but with singular Success: so that he feels that it is not too much to ask that he be forgiven for any degree of Intellectual Arrogance or Æsthetic Bigotry which may be detectable in the ensuing Pages.

The Author hopes that what is presented here as a sure Foundation upon which a genuine understanding of the Art can be developed will in fact prove so to be. The Book is divided into Three Parts: the "Basics", which a Child of 8 to 10 should easily be able to cope with (as they managed to in the 18th Century); the "Esoterics", which any quiet sensible Teenager ought to be able to follow; and the "Æsthetics", which could well be little more than a grievous wastage of Printing Ink and Paper; but which might otherwise prove to be of the greatest possible value to the Art of Practical Dialling.

Those who proceed seriously with this Art, whether Young or Old, Male or Female (and it is the Patience characteristic of the Female which is most suitable and requisite), will find that it is a most engrossing Study, and contains its own Rewards. Anything which militates against the prevalance of aggressive competitiveness in our Modern Society deserves at least to be warmly recommended.

Peter I. Drinkwater.
Shipston-on-Stour.
October 1984.

1

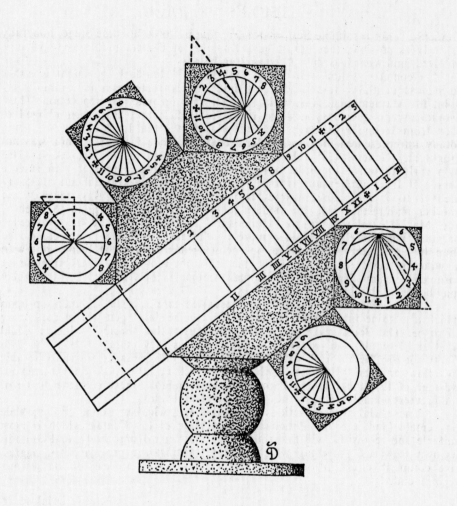

Basics

"Let the Dead awaken"

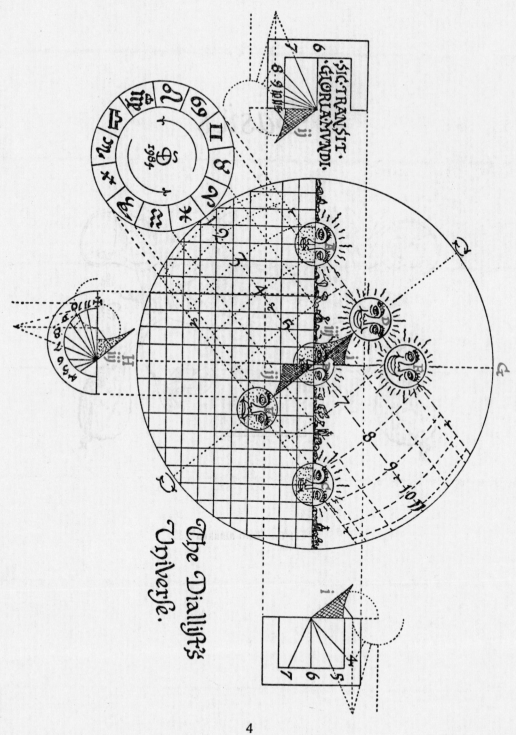

The Diallist's Univerſe.

SIC·TRANSIT·GLORIA·MUNDI.

4

BASICS

THE DIALLIST'S UNIVERSE

ACCORDING to actual Observation the Earth is a Sphære at the Centre of the Universe. Concentric with it, but so large that in comparison the whole Earth is a mere Mathematical Point, is the Great Sphære of the Heavens: the Cœlestial Sphære. Passing through the common Centre of both Sphæres is the Polar Line, the Axle-Tree of the Universe, around which each of the Heavenly Bodies completes one Revolution in about 24 Hours: the Sun goes round in *exactly* 24 Hours; and upon this Actuality the whole Art of Dialling is founded.

A double-sided Circle of Wood or Metal, marked into 24 Segments on each Side, with a long Pin running at Right Angles through the Centre, has the Capacity for "Telling the Time" by the Sun, in any Part of the Globe, and at any Season of the Year; providing (of course) that the Sun is above the Horizon, and is not obscured by Cloud. At the North Pole the Plate of this Dial is horizontal, with the long Pin (or *Gnomon*) pointing directly upwards to the Cœlestial Pole overhead. At this Point (as is the case at *any* Point on the Surface of the Earth) exactly half of the Cœlestial Sphære is visible; but the Cœlestial Æquator (or half-way Line between the two Poles), which at all other Latitudes cuts the Horizon due East and due West, is level with the Horizon. The Dial in this Position would tell the Hours "by the Sun" for the whole of those six Months when the Sun is on the North Side of the Æquator (21st March to 21st September, approx.), and be dark for the other six: but at the Poles all of the Meridian (or 12 o'Clock) Lines for the whole Globe meet at a single Point, and unless one of these be chosen at random there are no Hours to be told.

At any Point on the Terrestrial Equator (note the difference in Spelling) the *Gnomon* of this Dial is horizontal; and the Plate of the Dial is vertical, being exactly lined-up with the Cœlestial Æquator which passes overhead from a Point on the Horizon due East to another due West. The North and South Poles are also Points on the Horizon, directly opposite to each other, and are at Right Angles to the Æquatorial Crossings. The Dial in this Position will tell the Hours "by the Sun" for exactly 12 Hours every Day in the Year; for six Months on the North Side of the Dial Plate, and six Months on the South Side.

Wherever it is taken this simple Dial (called an Universal or Æquinoctial Dial) will tell the Time accurately, so long as it is correctly aligned with the Poles and Æquator; on one Side of the Dial Plate in Summer, and the other in Winter. And this is its limitation in our Latitudes, viz. that if it is set up high enough for the underside of the Plate to be visible, then the upper Side will be obscured from View; and if low enough for the upper Side to be visible, then the underside will be obscured.

5

BASICS

For this Reason it became usual to make the Plate of the Dial level (and concentric) not with the Cœlestial Æquator, but with the visible Horizon: the Gnomon continuing to point (as point it must) to the visible Pole. This type of Dial, the mystical "Horologiorum Eccentricorum" (Eccentric Dial) of late mediæval tradition or prosaïc Horizontal Dial of common parlance, will tell the Time "by the Sun" at any Season of the Year; the single calibrated Side of its Plate being exposed to the whole of the visible Hemisphære of the Heavens. It has the two disadvantages that (A) the Hour Lines upon it are unequally spaced; and (B) that there is a dichotomy between its requirement to be looked down upon, and its necessity to be high enough in the air to avoid the cast Shadows of surrounding Objects. It is essentially a private Instrument: for the Garden, not the Market Place.

For Public Use a Vertical Dial is necessary; since this can be set up at any Height from the ground. Such a Dial is exposed to only one half of the visible Hemisphære, one quarter of the whole Cœlestial Sphære. Since the Sun keeps to one half of the visible Hemisphære only during the Autumn and Winter it is necessary to provide at least two Vertical Dials to ensure the complete coverage of the longer Days of Spring and Summer.

The Problem, and its Solution, is expressed in the Orthographic Diagram, or modified Analemma, here presented: giving the View due East at Latitude 52° North.

The Diagram is largely self-explanatory. The Line 'G/H' is the due East Line; its solid portion being also the Dial Plate of a Vertical Dial facing due South, and another facing due North, seen edge-on and conjoined in one. The Gnomon 'i' is that of the North-facing Dial; the Gnomon 'ij' is that of the South-facing Dial; the Gnomon 'iij' is that of the Standard Horizontal Dial for the Latitude: the Dials themselves (rather the demi-Dials corresponding to the Times to be told when the Sun is East of the Meridian) being expressed in the Margins, with their embrionic Construction Details included. The Numbered Lines on the Central Sphære are those of the Solar Hours, as they would appear if they had been inscribed (by the Hand of GOD) upon the Surface of the Sphære itself, and were thus visible in the Sky. The Sun's anti-clockwise progress around the Circle of the Zodiac produces the varying Seasonal Tracks of the Sun across the Sphære, as here substantiated.

On Midsummer Day the Sun rises at a Quarter to Four in the Morning at Point 'A' on the Diagram, and registers on the North-facing Dial ('i'). By Six o'Clock it has reached Point 'D', but still registers on the North-facing Dial. At about Twenty past Seven it leaves the Plane of the North-facing Dial at Point 'F', and strikes the Plane of the South-facing Dial; registering upon the South-facing Dial until it is about Twenty to Five at Night, when it again enters

6

the Plane of the North-facing Dial and remains thereon until Sunset at about a Quarter past Eight at Night. A North-facing Dial, at this Latitude, must therefore contain the Hours of 4-7 a.m. (as shewn) and 5-8 p.m..

On the Day of the Spring or Autumnal Æquinox the Sun rises at exactly 6 a.m., Point 'B' on the Diagram, upon the Plane of the South-facing Dial, and registers on the South-facing Dial for the whole of the Day, until the Sun sets (again on the Plane of the Dial) at exactly 6 o'Clock at Night. The South-facing Dial must, therefore, contain all the Hours from 6 o'Clock in the Morning until 6 o'Clock at Night.

It must, however, be understood that the South-facing Dial will only register the Hours of 6 a.m. and 6 p.m. for two Days in the Year: the actual Days of the Æquinoxes; for (as we have seen) the Sun is registering these Hours on the North-facing Dial for the whole of the Spring and Summer; while in the Autumn and Winter it rises later, and sets earlier, than either of the Hours of Six. At Midwinter, at 6 o'Clock in the Morning, the Sun is at Point 'E', well below the Horizon; and does not rise until about a Quarter past Eight, at Point 'C' on the Diagram.

A Cubical Dial, to be mounted on a Cross or Column, needs (in its simplest form) to have these two Faces: North and South. To these Essentials are added (for symmetry) the standard East and West-facing Dials; which are calibrated so that the East-facing Dial gives the Hours from Sunrise to nearly Noon, and the West-facing Dial the Hours from a little after Noon to Sunset.

It needs to be understood that a correctly made Sundial records *Real Solar Time,* not the artificial 'Greenwich Mean Time' now in use; much less the 'British Summer Time' of yearly irritational occurence. This means that when the Sun is due South the Sundial will say 12 o'Clock and there will always be as many Hours of Daylight before Noon as after Noon: this is because Real Time is based on an equal division of a real Day of 24 Hours.

Mean Time, in contrast, is based upon an equal division of an whole Year; an entirely artificial division, which does not keep pace with the daily Real Time. During the Spring and Summer Quarters the difference between the two is never much more than 5 Minutes; but during the Autumnal Quarter 'G.M.T.' lags as much as 17 Minutes behind Real Solar Time, unnecessarily bringing forward the onset of the dark Winter Evenings; while during the Winter Quarter it becomes as much as 15 Minutes ahead of the Sundial, keeping the Mornings darker than they need to be, but brightening the Evenings. Another drawback to 'G.M.T.' is that it is based on Greenwich, in the South-East of the Country. Midday at Shipston is more than 6½ Minutes later than Midday at Greenwich; 7 Minutes later at Chipping Campden; and as much as 16 Minutes later in the more Westerly Parts of England (not taking account of

BASICS

Wales and Cornwall), with the Result that there is there a Discrepancy of more than Half an Hour between 'G.M.T.' and local Sundial Time during the Winter Quarter, which cannot but be commented upon by those who observe it.

There are several Methods for calibrating Sundials. The main Method hereafter to be described was first explained to English Readers by *Thomas Fale,* in his *Horologiographia* of 1593; it occurs earlier in published Continental Latin Treatises, and in Manuscripts dating back as far as the early 15th Century, soon after the general introduction into Europe of our Modern Method of dividing the Day. It is so very straight-forward, employing only the 'Ruler and Compasses' Geometry approved by the Ancient Greeks, as to be easily accessible to any, given the requisite desire to acquire the Skills involved.

BASICS

Præterit iſte dies:
neſcitur origo ſecundi:
An labor, an requies:
ſic tranſit fabula mundi.

This present day so very swiftly goes;
The next, in rising, what it holds for us,
We know not; whether toil or rest: 'tis thus
This world's vain story hastens to its close.
(H.W.)

9

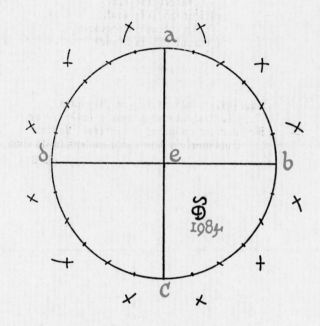

TO DIVIDE A CIRCLE INTO HOURS, HALVES, QUARTERS, MINUTES,
MILEWAYS AND DEGREES

BASICS

TO DIVIDE A CIRCLE INTO HOURS, HALVES, QUARTERS, MINUTES, MILEWAYS, AND DEGREES.

THERE are 24 Hours in a Full Circle.

'abcd' is a Circle on Centre 'e', quartered by Diameters 'ac' and 'db' at Right Angles to each other. From 'a', 'b', 'c', and 'd', set out the Radius ('ae') along the Circumference in each Direction: this gives 12 equal divisions. Bisect each of these in the usual geometrical manner to yield the 24 Equal Hours of the Full Circle.

To make Half Hours bisect each division again.

To make Quarters bisect again.

To make Intervals of 5° pace each Hour, with Dividers, into 3 equal parts: these are called "Mileways", because in the Time represented (20 Minutes) one is reckoned (on average) to walk a Mile.

With a little Practice one can divide a Mileway into 5 individual Degrees by eye.

11

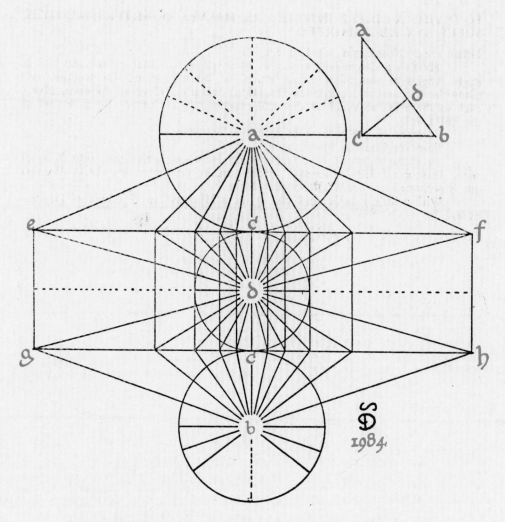

THE RECTIFICATORY AND FUNDAMENTAL DIAGRAM.

12

BASICS

THE RECTIFICATORY AND FUNDAMENTAL DIAGRAM.

THE "Rectificatory" is a Right Angled Triangle 'abc'; of which 'acb' is the Right Angle, 'abc' the Latitude of the Place for which the Sundials are required to be made, and 'cab' the complement of this Angle.

All Diagrams in this Book are drawn for 52° North, because the Author lives only fractionally to the North of the 52nd Parallel of Latitude.

Latitudes of any Place in England can be obtained from a Standard Ordnance Survey Map.

'cd' is a Line extending from 'c', to meet Line 'ab' at Right Angles. Early Diallists actually *made* the Rectificatory in the form of a Set Square, fitted with a Plumb-line for levelling, and used it as a base for all of their Constructions.

'cd' is taken as the Radius of the Æquinoctial, or equally divided Circle of Hours. Draw this Circle on Centre 'd', and divide it into Hours as taught. Establish Tangent Lines 'ef' and 'gh' at Right Angles to the Prime Vertical. From the Centre 'd' draw the Equal Hour Lines straight through to meet these Tangent Lines.

Early Diallists actually *made* the Æquinoctial Circle, usually with a Sector missing from the Top so that it could be slipped over a Rod propped up at the correct Angle with the aid of the Rectificatory, and employed String (fixed into little holes in the Circle) to extend the Hour Lines to meet the Plane on which the Dial was required to be drawn.

'bc' is the Radius of the Horizontal Plane proportionate to the Æquinoctial Radius at the Latitude in question. Set 'bc' along an extension of the Prime Vertical downwards, and draw the Horizontal Circle with this Radius to the Tangent Line 'gh'. The Hour Lines for the Horizontal Plane go from the Points on this Tangent Line to the Centre 'b', and through this Centre where so indicated (see Pages 21 & 39). The Gnomon for this Plane stands over 'cb' at the Angle 'abc': it can be a Solid Triangle, an Angled Rod or Wire, or a Strutted String. 'c', the Midday Meridian (12 noon) points due North, with the Morning Hours (11 a.m. back to 4 a.m.) to the West, and the Afternoon Hours (1 p.m. on to 8 p.m.) to the East.

'ac' is the Radius of the corresponding Vertical Planes facing due South and due North. Set 'ac' along an extension of the Prime Vertical upwards, and draw the Vertical Circle with this Radius to the Tangent Line 'ef'. The Hour Lines for the Vertical Planes go from the Points on this Tangent Line to the Centre 'a'. For the due South Plane they do not rise above the Horizontal Diameter of the circle. The Gnomon for the due South Vertical Plane stands over 'ac' at the Angle 'bac', pointing downwards. The Midday Meridian also points downwards, with the Morning Hours (11 a.m. back to 6 a.m.) to the West, and the Afternoon Hours (1 p.m. on to 6 p.m.) to the East. The Gnomon for the due North Vertical Plane points upwards; as does its Meridian, in this case Midnight. The due North Vertical Plane has Hour Lines drawn right through the Centre 'a', as shewn by the dotted lines: only the Hours 4 a.m.-8 a.m. are included to the West, and 4 p.m.-8 p.m. to the East.

The Vertical Parallels from 'eg' to 'fh' form the Hour Lines upon all *direct* Planes parallel to the Axis of the Earth at any Latitude. The Shadow on all of these Planes is cast by the Tip of a Rod with the length 'cd', set upright in 'd'; or by a Straight-edge parallel to 'cdc', at a distance of 'cd' from the Plane. The selection & numbering of Hour Lines for these Planes is various, as explained hereafter.

The Fundamental Diagram, if understood aright, provides the general Key to all *Direct* Dialling; that which follows is a series of exemplifications in particular cases.

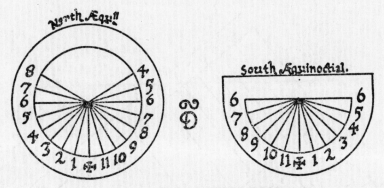

THE ÆQUINOCTIAL DIAL.

THE ÆQUINOCTIAL DIAL.

ESSENTIALLY a double-sided Circle divided into 24 equal parts, for the 24 Hours. The Upper (Northern) Face can retain only the Hours of the Longest Day at the Latitude of erection; but, if adjustable, needs a full Circle of Hours. The Lower Face needs only those from 6 a.m. to 6 p.m. The Gnomon for this Dial is a plain Rod set up perpendicularly in its Centre, and goes straight through the Dial, standing parallel to the Axis of the Earth; whilst the Dial itself lies in the Plane of the Æquator, hence its name. It is usual to make this Dial Adjustable to any Latitude (rather any *Northern* Latitude), *and* portable, as may easily be done. If it can be pivoted so as to turn right round (which I have never seen) it would be truly "universal" or usable anywhere in the World: in such an instance both North & South Faces would need to be full Circles.

It is an Horizontal Dial (but a meaningless one) at the Poles, and a Vertical Dial at the Æquator; where both North & South Faces could only be marked from 6 a.m. to 6 p.m., and would each serve for a full 6 Months of the Year.

This is the Dial with which all others can be calibrated.

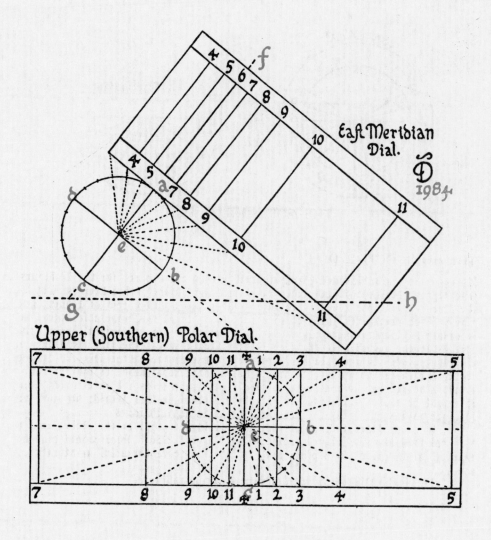

East Meridian Dial.

Upper (Southern) Polar Dial.

THE MERIDIAN AND POLAR DIALS.

16

BASICS

THE MERIDIAN AND POLAR DIALS.

THESE Dials are all based on a simple Tangent Scale: the Gnomon (in each instance) being either the Straight-edge shewn, or the Tip of a Rod of the length 'ae'; the Prime Radius of the Tangent Scale.

The Meridian Dials are the Vertical due East and due West Dials everywhere, their Latitude being the Angle 'fgh', and the only required adjustment to the Construction for any place on Earth: except for the expected amendation of Sunrise/Sunset Times, specific to the Latitude, for the Longest Day.

The Polar Dial is the Horizontal Dial at the Æquator, and is so called because it lies in the Polar Plane; it is a Vertical Dial without meaning at the Poles, where there is no Time. When made for other Places it has a complementary Lower (or "Northern") Face; which is, in practice, rarely constructed.

Meridian & Polar Dials can be effectively made "Universal", and combined with the Æquinoctial Dial, as shewn.

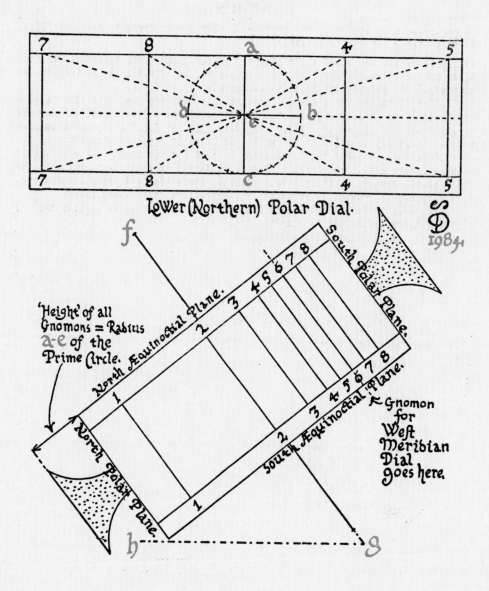

Lower (Northern) Polar Dial.

ED
1984

'Height' of all
Gnomons = Radius
a-e of the
Prime Circle.

North Æquinoctial Plane.

1 2 3 4 5 6 7 8

South Polar Plane.

f

North Polar Plane.

1

South Æquinoctial Plane.

1 2 3 4 5 6 7 8

Gnomon
for
West
Meridian
Dial
goes here.

h

g

18

BASICS

Hie kurtze frist:
Dort ewig ist.

Bedenck das end/
Zeit laufft behend.

Die stund hin fehrt/
Nicht wider kehrt

Die zeit verschwindt
Ehe mans befindt.

Here short respite:
There long repose.

Think on the end/
Time swiftly goes.

The hour doth pass/
Returning never

The time moves on
Unnoticed ever.

(H.W.)

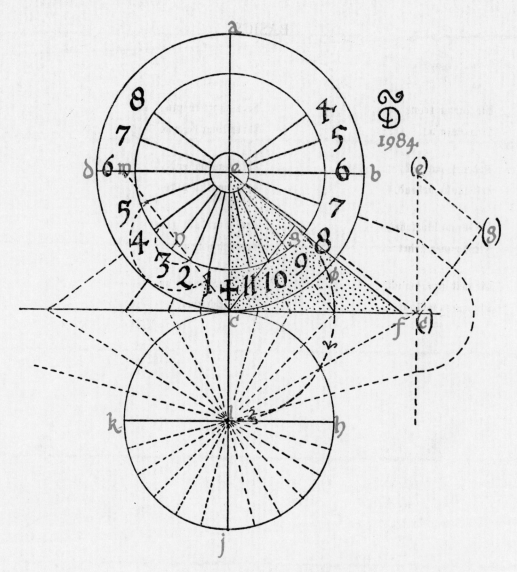

THE HORIZONTAL DIAL.

20

BASICS

THE HORIZONTAL DIAL.

THE Horizontal Circle is 'abcd'. 'ceφ' is the requisite Latitude. 'gc' is the Æquinoctial Radius. 'cl' = 'gc'. 'chjk' is the Æquinoctial Circle, equally divided into 24 Hours, with lines from the Centre 'l' to the common Tangent Line. The Hours of the Horizontal Plane run from the Points on the Tangent Line to the Centre 'e'. 'cef' is the Gnomon.

To avoid the awkwardness of a long Tangent Line, either:

(A)   Mark 'em' equal to 'gc'. 'mc' crosses 3 (or 9) o'Clock at 'n'. With 'n' as a Centre, mark up the Points where the Hour-lines cross 'nm', symmetrically with their established crossings of 'nc' as indicated, using the Compasses. This applies for all Hour Lines, Half-hour Lines, and Quarters. This Method can only be used on Direct Dials (i.e. those which have an Hour Line directly under the Gnomon, in the position of the "sub-style").

or:

(B)   Draw a Perpendicular through the Tangent Line at any random distance from 'c'; reproduce Triangle 'ceg'. On '(c)' as a Centre, using the Compasses, transfer the distances as shewn, and draw cross lines through the Triangle '(c)(e)(g)' parallel to '(e)(g)'. The remaining Hour Lines extend from the Points thus established along '(c)(e)' to the Centre 'e'. This Method works for Indirect (i.e. Declining etc.) Dials as well as for Direct.

The Horizontal Dial is the standard "Garden Dial", and as such is well known to all. The "Back Hours", 4 a.m.-6 a.m. and 6 p.m.-8 p.m., are requisite at our Latitudes.

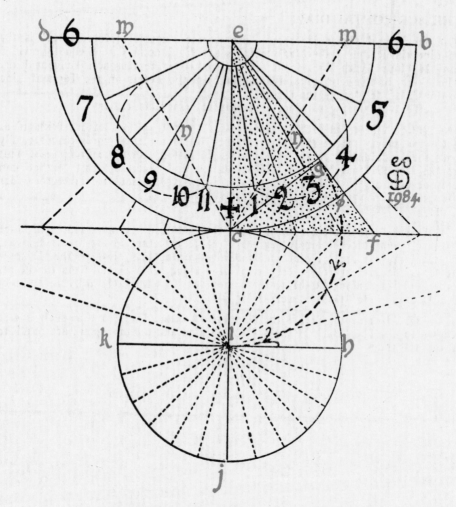

THE VERTICAL DIRECT SOUTH DIAL.

22

BASICS

THE VERTICAL DIRECT SOUTH DIAL.

THIS is a Dial on a Wall (or Dialstone) facing due South, and is drawn as an Horizontal Dial except that the Latitude is 'beϕ' (and 'efc'), while the Gnomon is 'cef'. There are no Hours to be marked above the 6 o'Clock Line, and the Shadow will only rise that high (but hardly then) on the actual Days of the Æquinoxes.

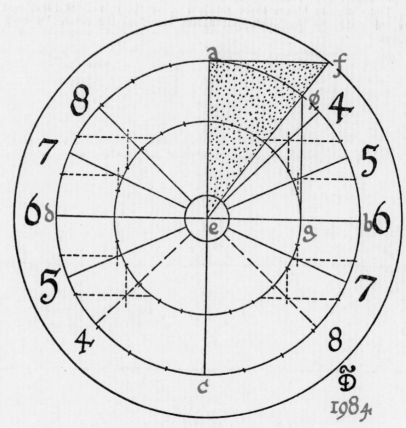

THE VERTICAL DIRECT NORTH DIAL.

24

BASICS

THE VERTICAL DIRECT NORTH DIAL.

THIS IS the same as a Direct South Dial but on a North facing Wall (or Dial-stone), inverted, and with a different Selection of Hour Lines, including "Back Hours". The Hours 4 a.m.-8 a.m. & 4 p.m.-8 p.m. are traditional, but only 4 a.m.-7 a.m. & 5 p.m.-8 p.m. are requisite for our Latitudes. The Gnomon must point upwards to the visible Pole.

As a change from the usual Tangent-Line Method it may (as may most other Dials) be drawn by Orthography: 'φeb' is the Latitude, 'aef' the Gnomon. 'abcd' is the Prime Circle. 'φg' is parallel to 'ae'. 'eg' is the Radius of the Minor Axis Circle. Each Circle is divided into 24 Hours (with the Halves and Quarters). The (dotted) Horizontal Parallels spring from the Hour Points (etc.) of the Prime Circle. The shorter Vertical Parallels spring from the corresponding Points on the Minor Axis Circle. The Hour Lines go through the Points where the corresponding Horizontal and Vertical Lines cross.

The Problem of constructing Dials on Walls *not* facing due East, West, South, or North, requires the Understanding to be taxed with additional Geometry. The first Generation of Diallists (15th to early 16th Centuries) avoided the Problem by the use of the Rectificatory & Circle, working empirically "in the round" as it were. The Edge of the Gnomon casting the Shadow remains in line with the Axis of the Earth, and the Noon Shadow is always perpendicular. This *could* be achieved by "folding" the Gnomon of a Direct South Dial towards the South along its own Sub-style until it is in line with the Meridian (vice-versa for a North Dial); but it makes more sense to have the Gnomon standing square to the Wall, and this is worked out by simple sectioning of an arbitrary Horizontal Plane cutting the whole Construction.

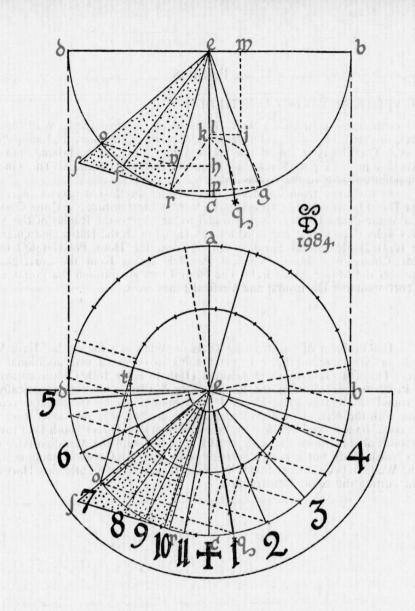

VERTICAL DIALS DECLINING FROM DUE SOUTH. *EXAMPLE: DEC-LINING 21° EASTWARD.*

26

BASICS

VERTICAL DIALS DECLINING FROM DUE SOUTH. *EXAMPLE: DEC-LINING 21° EASTWARD.*

LAY out the Latitude 'def', and the Declination (or slew of the Wall from due South) 'ceg'. 'fh' is parallel to 'db'. Set out 'ej' equal to 'fh'. Then: 'hn' equals 'jl'; and 'no' equals 'jm'. 'f(o)er' is the Gnomon in its proper location. This becomes understandable when one considers the Quadrant 'ceb' as representing an horizontal Section through 'fh' looking upwards from underneath. One now has sufficient Requisites to draw the Hour Lines with a Tangent and Circle.‖

To proceed by Pure Orthography, however, make 'rp' parallel to 'fh'. Then: 'pk' equals 'pg'; and 'rq' equals 'rk'. The Angle 'req' is the "Inclination of Meridians", or Angle between the Sub-style 're', and the vertical Noon Line 'ec', in the Plane of the Æquator.

Then thus:

Quarter the Prime Circle respective to Point 'r'; 'ot' being parallel to 're', 'te' is the Radius of the Minor Axis Circle. Quarter both Circles respective to Point 'q', and divide into 24 Hours. Parallels *square* to 'er' spring from the Hour Points on the Prime Circle; Parallels *to* 'er' spring from the Points on the Minor Axis Circle. Mark where each corresponding Pair of Lines crosses: the Hour Lines go through these Points from the Centre 'e', and include the Meridian, which must correspond with the Vertical 'ec' already drawn or the Work has been done wrongly.

To draw with a Tangent Line, proceed from '‖' thus:

Draw the Tangent Line through 'r'; prolonging 'er' below it. Prolong 'ec' to 'z' (under Noon Cross on Diagram). 'rx' equals 'vr' (which is square to 'fe'). Join 'x' and 'z'. 'rxz' is the "Inclination of Meridians", and the Æquinoctial Circle (to Radius 'xr') is drawn on Point 'x'. Quarter the Circle respective to 'xz', and divide into 24 Hours. Projection from Centre 'x' to the Tangent Line, and from thence to Centre 'e', should by now be self evident. The Method of foreshortening indicated, and already taught on Page: 21, is most needful in the drawing of this Type of Dial by the Tangent-line Method: due to the fact of the "Difference of Meridians" some of the outermost Projection Lines would otherwise meet the Tangent Line at a vast distance out, especially if it is wished to include Halves and Quarters. Note that a separate repetition of the Triangle 'rve' is required for each side of the Dial.

N.B. that on Dials Declining *West* the Sub-style "disflects" in the opposite Direction to that shewn here. Note also that the Spread of Hour Lines on *all* Vertical South-East and South-West Decliners is limited by the Horizontal 'deb'. Some old Diallists included Lines occuring in the Sector '(e)ed'; but the Shadow cast by any Horizontal, or downward-pointing, Gnomon can *never* rise

BASICS

that high! Hour Lines presenting themselves in that Sector should be prolonged straight through the Centre 'e' to the other Side of the Dial, and the original Generation deleted. Hour Lines presenting themselves *below* the Horizontal, but representing Times *before* Sunrise, or *after* Sunset, on the Longest Day ought by no means to be included, except (by tradition) the first of them. An empty Sector will therefore emerge, above the Gnomon, as the Dial's Declination from due South increases.

"Great Decliners" or Dials facing very nearly due East or due West present a special Problem, as they cannot effectively be drawn within a Semicircle. However, at the expense of the Noon Line, they can effectively be constructed.

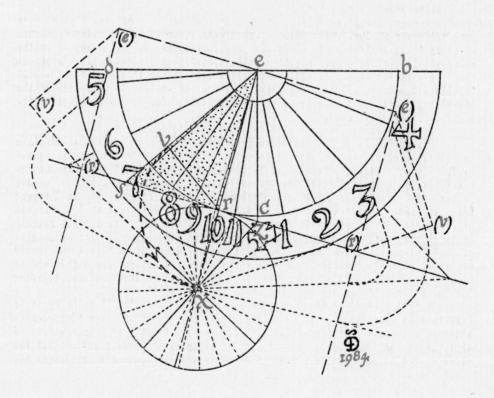

BASICS

NORTH DECLINERS.

THESE are constructed (but rarely) in the same Manner as a South Decliner; but their Gnomons point *upwards* to the Visible Pole, and they have "Back Hours" (like an Horizontal Dial), with an empty Sector around the Meridian of Midnight. Like South Decliners their Gnomons "disflect" in the opposite Direction to their Declination.

Der heutig tag also hin schleicht/
Wer weisz wer den morgend erreicht?

Today's today doth steal away/
Who knows who'll reach the morrow?

(H.W.)

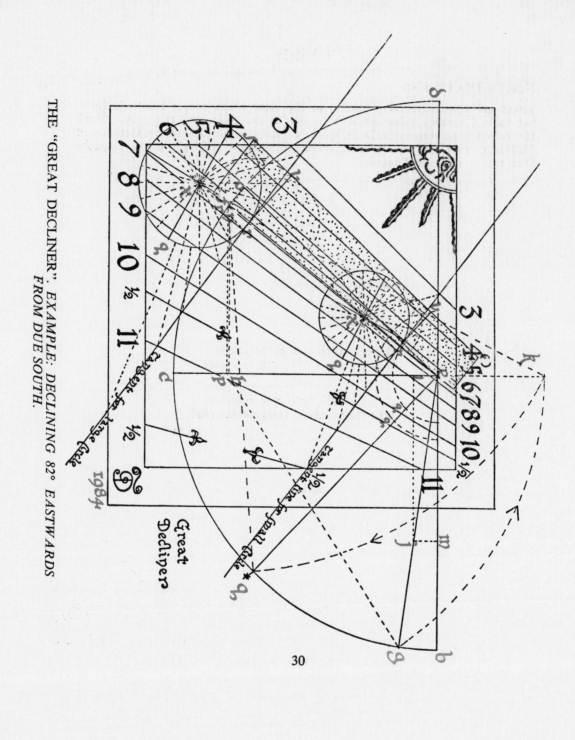

THE "GREAT DECLINER". EXAMPLE: DECLINING 82° EASTWARDS FROM DUE SOUTH.

30

BASICS

THE "GREAT DECLINER". *EXAMPLE: DECLINING 82° EASTWARDS FROM DUE SOUTH.*

LAY out the First Diagram on Page: 26 for the given Situation. Some of the Points will present themselves in unexpected Places (i.e. 'k' *above* the Centre 'e'), others will appear excessively close together: this is all correct, being caused by the extreme Situation of the proposed Dial.

The Resultant Gnomic Triangle 'oer' is far too narrow to be made use of. Draw therefore 's(e)' parallel to 'oe' at any selected distance. Draw the *two* Tangent Lines shewn (at any chosen position) at Right Angles to the Substyle Line 'er'; and the two Æquinoctial Radii 'vr' & 'vr' at Right Angles to the Style 's(e)'. 'rx' equals its respective 'vr' in each instance. Draw the *two* Æquinoctial Circles; with semi-circles to the same Radii on Centre 'e', so that the Chord of the Angle 'req' can be taken off to the correct radius. Set this Angle from 'r' to 'q' in each instance. Quarter the two Circles from point 'q', and divide into 24 Hours. Project to the Tangent Lines, and connect the corresponding Points on the two Lines to make the Hour Lines. Half Hours are put in as shewn, *only* where there is room for them. The Position, and Shape, of the confining Border is quite arbitrary; and according to personal taste.

"Great Decliners" declining to the West require the Construction to be reversed. Those declining from the North require its inversion, so that the broad Part of the Gnomon is uppermost, and pointing to the visible Pole.

31

BASICS

RECLINING AND INCLINING DIALS.

UNLESS one wishes to emulate the "Philosophical Diallists" of the Renaissance in the Construction of Dials on each Facet of any given Geometrical Solid, or Polyhedron, the understanding of such Dials is a most unprofitable Study. This being said, Dials *reclining* from the Horizontal to the East or West may be drawn, with common-sense modifications, in the same Manner as Vertical Decliners; whilst those *reclining* to the North or South are even simpler to produce: two varieties of the latter, the Æquinoctial and Polar Dials, have already been fully explained.

Dials which both decline and recline (or incline) present one with undreamed-of Complexities, requiring as much additional Knowledge as would fill a Book three times the Length of this one. Nor did the Old Diallists themselves quite get it right; with the Honourable Exception of the Englishman *Thomas Fale* most of their "Methods" are an absolute confusion from beginning to end. Many have tried but most have failed, and "Silence is most noble in the end".

This *again* being said, the present Author has wasted a great deal of Time in the study of these Complexities, and will cheerfully provide Copies (D.V.) to any interested Parties who apply to him; but might eventually be foolish enough to produce a Book upon this very Subject, if Partners to his Folly can effectively be located.

BASICS

Adfcriptum horologio Curię Heidelbergenfis.

Merck ſtund und tag/   Wie ſie ſo gach
Fahren dahin/   Und wir mit in,
Drumb ſihe zu   Bey zeiten nu/
Dasz dich ja find   Der tod geſchwind
Zur fahrt bereit/   Der dein nicht beit/
Der eilend Bott   Geſand von Gott.
Wie der macht dann/   So muszt dus han/
Freud oder leid   In ewigkeit.

Inscribed on a Sundial on the Town Hall of Heidelberg.

Mark hour and day/   O how they go
Fast slipping by/   And we also,
Therefore beware   Whilst with time grac'd/
Be found prepar'd   When death in haste
Bids journey hence/   He won't exempt
That speedy Envoy   God hath sent.
What he decides/   So must it be/
Eternal joy   Or misery.

(H.W.)

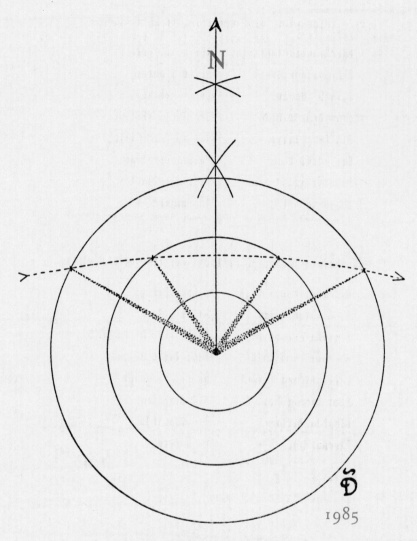

TO LOCATE TRUE NORTH.

34

BASICS

TO LOCATE TRUE NORTH.

PREPARE a flat Board with any number of Concentric Circles drawn on it, and set up a Thin Rod, absolutely perpendicular, in the common Centre. The Board is itself Square; so that, when it is laid down absolutely plumb Horizontal, one of the Edges may be set against any Wall whose Declination (or Slew from due South, or due North) is required to be found. Starting in the Morning (or in the Evening in Summer for a generally North-facing Situation) the Points where the Shadow of the Tip of the Thin Rod crosses any of the Concentric Circles is to be marked on the Board; and again when it re-crosses them in the Afternoon (or Morning of the Following Day in Summer for generally North-facing Situations). When the Angle between each Pair of Marks is bisected the Meridian (or due North/due South Line), which should be the same in each instance, is located; together with the Declination of any Wall against which the Board has been set. This Method works best at the Summer or Winter Solstice, when the Sun's own Declination is most constant, and worst at the Spring or Autumn Æquinox, when its rate of change is most rapid.

If this Construction is performed upon the Surface of a Wall, with a True Horizontal Rod projecting therefrom, what is found is *not* the "Meridian of the Place" but the "Meridian of the Plane" or Sub-stylar Line. If this is located thus, and the Latitude of the Place is known, the remaining Requisites (the Declination of the Wall and the Angle of the Gnomon) are easily established by an understanding of the Geometrical Construction already taught on Page: 27. Remember that the drawn Sine of the Co-latitude is the Hypotenuse of a Right-Angled Triangle, of which one other Side is the Portion cut off this Sine by the crossing of the Sub-stylar Line, and the remaining Side the Sine of the Angle of the Gnomon to be established.

So, in Dialling, as in Geometrical Construction generally, all things are inter-related, and can be used to establish each other.

This Method for finding the Meridian has been known and used from the earliest Times to the near Present; occuring as it does in the Writings of *Vitruvius, Bede,* and *Chaucer.*

Be very wary of using the Magnetic Compass to establish True North. Universal Reliance upon this Device in the 16th Century, and earlier, resulted in the construction of many "False Dials" which had to be corrected or replaced within a Century. The position of Magnetic North is completely mobile. Prior to 1600 it lay generally East of True North, decreasing all the while; after 1600 it lay to the West, as it does still, to the extent of a good 7½° from where we are. Some Compass Makers made a contemporary correction (which soon became wrong), others made none. Some Diallists applied a correction (frequently an erroneous one), others did not. Nobody took notice of the effects of "local Iron" in deflecting the Needle from the true. Etc..

As a matter of interest, the Mediæval Compass Directions were not, of course, the modern English N(orth), S(outh), E(ast), and W(est), but the Latin S(eptentrio), M(eridies), Or(iens), and Oc(cidens).

Esoterics

"... thinking GOD's Thoughts after Him ..."

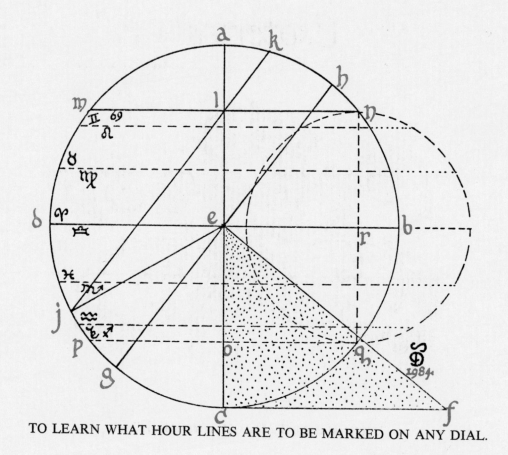

TO LEARN WHAT HOUR LINES ARE TO BE MARKED ON ANY DIAL.

38

ESOTERICS

TO LEARN WHAT HOUR LINES ARE TO BE MARKED ON ANY DIAL.

THERE are *three* Factors to be taken into account:

1) THE SHADOW-CASTING CAPACITY OF THE GNOMON ON THE PLANE,
2) THE HORIZONTAL FACTOR,
3) THE TIMES OF SUNRISE AND SUNSET ON THE LONGEST DAY.

Early Diallists often regarded only the first of these, with strange results; it is found thus:

'abcd' is the Prime Dial Circle; 'cef' the Gnomon. 'geh' makes Right Angles with 'ef'. Angle 'jeg' is 23½°, and 'jlk' is parallel to 'geh'. 'mln' is parallel to 'deb'. Hour Lines potentially passing through the Segment 'lman' are not proper to the Plane: they can never take the Shadow on any day of the Year when the Dial is properly aligned. This Construction is all that is required to determine the spread of Hour Lines on an Horizontal Dial, as also the Times of Sunrise and Sunset on the Longest Day.

All *Due South Dials* and *South Decliners* are limited by the *Horizontal Factor* alone, with allowance for factor (3) in extreme (but see below for "great") decliners. The *Horizontal Factor* is no more than an Horizontal Line drawn through the Root of the Gnomon (Point 'e' in most of the Drawing) cutting out anything which might otherwise be drawn above it. Being such a simple device it is surprising that the earliest Diallists did not have a better understanding of it.

All Due North Dials and North Decliners are limited by Factors (1) and (3) only. The Horizontal Factor applied upside-down *does not* apply to them, although many early Diallists imagined that it did.

Special cases (Æquinoctial, Polar, and Meridian Dials) have already been satisfactorarily dealt with. Great Decliners never achieve the Meridian, and are further limited by the Times of Sunrise and Sunset on the Longest Day, with the allowance of an extra Hour "for shew".

If it is required to know the Limits of the Shadow-casting Capacity of the Gnomon on the Plane on *other Days* but the Longest, all else being considered, draw 'nrq' parallel to 'aleoc', draw a Circle with radius 'rn' on Point 'r', divide it equally into 12 Parts, and extend the Parallels, as shewn, across the Prime Dial Circle, cutting off increasingly large Segments of Time.

The Signs of the Zodiac, as drawn here for the Horizontal Plane (on Vertical South Planes Capricorn is at the top, and Cancer at the bottom), correspond with the Dates of the Cusps of the Signs as well-known to Students of Popular Astrology. A few early Diallists used the *minimum,* rather than the *maximum,* Capacity for the limitation of Hour Lines on Vertical Direct South Planes, as applies at Midsummer; an arrangement which appears on the Small Sundial included by Albrecht Dürer in his famous Engraving *Melancholia,* where it may well be studied; although he has an inexplicable hiatus in the numbering of the Hours.

THE Preceding is a specialized application of the Manæus (from the lost Greek Mηναîos), or "Circle of the Months", which formed part of the Analemmatic (or Orthographic) Spherical Projection of the Ancients, being remembered and used by the Diallists of the later Middle Ages. It determines, most precisely, the Sun's Seasonal Declination (as an Angular Distance) from the Æquator. With its aid several interesting Types of Sundial may be constructed.

ESOTERICS

Zeit feiret nicht/ sey fertig/
Des tods all stund gewertig.

Time waiteth not/ be always ready/
Death's presence at all hours is steady.

(H.W.)

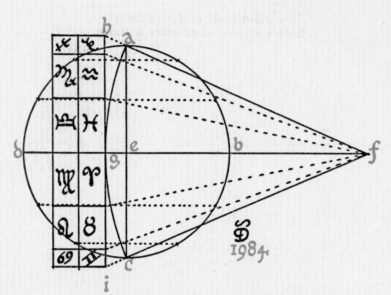

THE MANÆUS (ΜΗΝΑΙΟΣ) AND TRIGON.

ESOTERICS

THE MANÆUS (MHNAIOΣ) AND TRIGON.

To make a Tangent Scale of Solar Declination to a Specific Radius ('fg').

SET out Angles 'afg' and 'gfc' at 23½° each (the Precise Figure, nearly 24° in Classical Times, is now a little less than 23½°), and draw the Arc 'agc' on the Centre 'f'. Connect 'a' and 'c' with the Chord-line 'aec'. Upon the Centre 'e' draw the "Circle of the Months", divided into 12 Equal Parts as shewn, with its Horizontal Parallels to 'dgebf'. Draw the requisite Angles from Centre 'f', through the Points marked by these Parallels on the Arc 'agc' (many early Diallists, *in grave error,* used instead the Points on the Chord-line 'aec'!) to the Tangent-line 'hgi', parallel to 'aec'. The Signs of the Zodiac:

"The *Ram,* the *Bull,* the *Heavenly Twins,*
And, next the *Crab,* the *Lion* shines,
The *Virgin,* and the *Scales;*
The *Scorpion, Archer,* and *Sea-goat,*
The *Man-who-pours-the-water-out,*
And the *Fishes* with glittering tails"

or "Aries, Taurus, Gemini, Cancer, Leo, Virgo, Libra, Scorpio, Sagittarius, Capricorn, Aquarius, and Pisces" as they are now generally well known, are here marked with the Tropic of Capricorn at the Top, and that of Cancer at the Bottom, as required for some purposes; for others they are required to be the other way round. If it is required to mark the Decanates (or intervals of 10° Approximating to 10 Days) of the Zodiac, this is done by stepping each of the 12 Divisions of the Manæus into 3 equal parts, and projecting: this is usual, divisions of less than 10° are not.

The Triangle 'hfi', with its Scale and internal markings radiating from 'f', is the "Trigon"; which the early Diallists actually made as an Instrument. It was incribed upon a Rectangular Board, 'hi' forming one edge, and a Parallel to 'hi', through 'f', the other. Holes were drilled along 'hi', into which Strings were knotted to prolong into space the Lines radiating from 'f'. The Edge containing Point 'f' was "hinged" to the Edge of a Sundial Gnomon, so that 'f' was placed against a Notch or "Nodus" already cut therein, and the Strings stretched out to make contact with the Dial-plane; marking out, in each instance, a line of points forming one of a Series of Hyperbolic Curves to be tracked by the Shadow of the Nodus at the appropriate "Season" of the Year. A more rational way of performing the same is given hereafter. The "Trigon", as an Instrument, persisted longer than the similarly envisaged "Dialling Circle" mentioned on Page 12a. String was known as an Engineering tool well before the Days of Heath-Robinson!

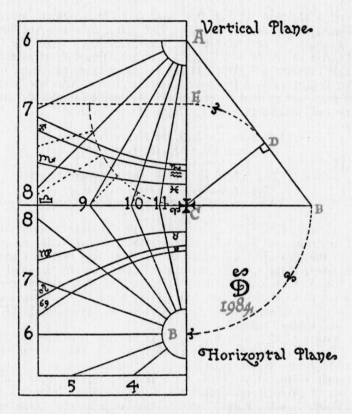

TO MARK THE "PARALLELS OF SOLAR DECLINATION" ON ANY DIAL. *EXAMPLE: A DIPTYCH DIAL* (HORIZONTAL AND VERTICAL (SOUTH-FACING) PLANES).

ESOTERICS

TO MARK THE "PARALLELS OF SOLAR DECLINATION" ON ANY DIAL. *EXAMPLE: A DIPTYCH DIAL* (HORIZONTAL AND VERTICAL (SOUTH-FACING) PLANES).

THE Dial is designed so that the two Planes meet at a common Æquinoctial. 'ABC' is the Gnomic Triangle. 'CD' is the Æquinoctial Radius; with 'E' as the Point of projection.

On a separate Diagram set out 'ADB' as a Vertical, with an extended Horizontal from 'D'. Take the Lengths of the individual Secants of the Æquinoctial Hours, from 'E' to the common Tangent Line, and set them out from point 'D' on the second Diagram along the Horizontal. Draw the Hour Lines to (in other cases through) these Points from 'A' and 'B'.

Construct the Manæus as shewn.

Take the requisite distances, to the individual "Parallels", along each Hour Line on the second Diagram from Points 'A' and 'B', and transfer them to the corresponding Hour Lines on the Dial. The Hyperbolic Parallels will then emerge.

To mark the *Horizon* onto the Dial Plate of any *Vertical* Dial; draw a line from the Nodus to meet the Sub-stylar Line at Right Angles. The true Horizon Line passes through this specific Point on the Sub-style, and is as precisely horizontal as one would expect! On Declining Dials the Horizon, Æquinoctial and 6 o'Clock Lines will cross at one single Point, unless the Work has been done wrongly.

On the East and West facing Dials, which have no "Centre 'A' or 'B'" (as also Dials on the Polar Plane), the Hour Lines on the Second Diagram stand at Right Angles to the Horizontal Extension from 'D'.

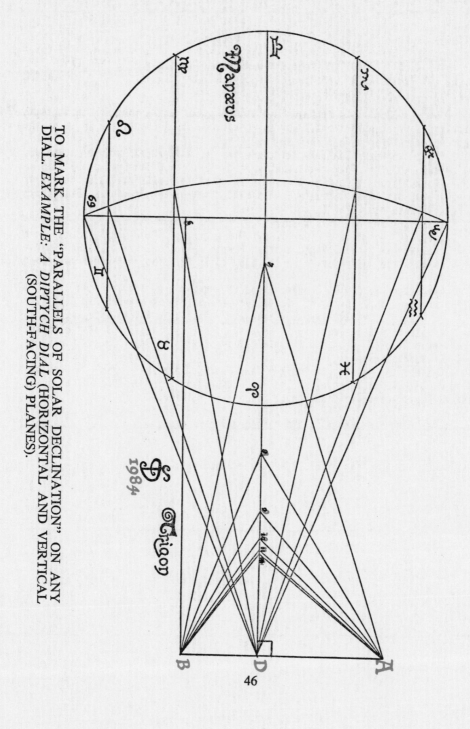

TO MARK THE "PARALLELS OF SOLAR DECLINATION" ON ANY DIAL. EXAMPLE: A DIPTYCH DIAL (HORIZONTAL AND VERTICAL (SOUTH-FACING) PLANES).

46

ESOTERICS

Der schadt der stang dir bed[e]ut
Die glyche stund zu aller zyt
Tags lenge und sonnen zeychen
Thut der knopff mit synem schatten erreychen.

The shadow of the style hath power
To tell the time at any hour
The day's lengths and the solar signs
These things the little nodus with its shade divines.

(S.M.)

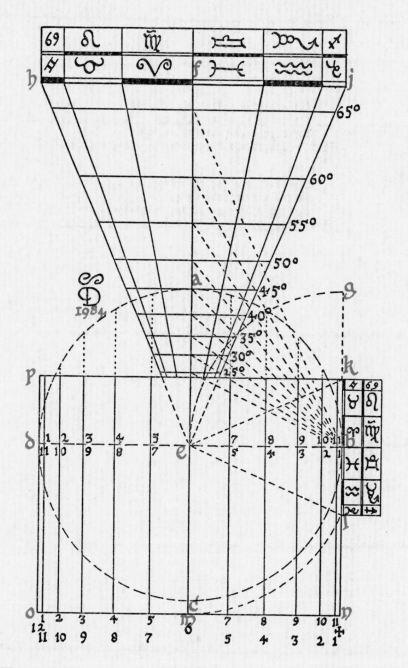

THE REGIOMONTANUS, OR GENERAL QUADRANGULAR, DIAL.

48

ESOTERICS

THE REGIOMONTANUS, OR GENERAL QUADRANGULAR, DIAL.

TRADITIONALLY ascribed to *Johann Müller,* called "Regiomontanus" (The Man from the Royal Mountain), its supposed Discoverer: lived 1436-76, being poisoned in Rome. He was a Native of *Königsberg* ("King's Mountain") hence the Name.

'abcd' is the Prime Circle on Centre 'e', quartered, with the Vertical Diameter extended upwards towards Infinity. On Centre 'b', using Quadrant 'eg', mark a Tangent Scale of Degrees of Latitude up this extended Diameter from 'e'; it has no natural upper limit; although the Arctic Circle (66½°) is traditional, this may well be exceeded. It should also be continued right down to 'e', the Æquator: this Detail being omitted from the Diagram for clarity. Angles 'hef', 'fej', 'keb', and 'bel' all equal 23½°. Draw the Horizontal Parallels of Latitude within the larger Trigon. Using the Principle of the Manæus, construct the two Tangent Scales of Declination along 'hj' and 'kl', and extend the Lines of Declination from the Points on 'hj' to 'e' so that the Trigon 'hej' is completely calibrated. Mark in the Signs of the Zodiac as shewn. 'em' equalling 'el', draw the Table 'pkno'. Divide the Prime Circle into 24 Hours, and draw the Vertical Parallels throughout the Table, for the Hour Lines, with subdivisions for the Halves as possible. Decanates to the Declination Scales, and individual Degrees (so far as is possible) to the Latitude Scale, should also be provided; together with the Hour Numbers (reserving the Cross ☩ for Noon, with 12 for Midnight) in either or both of the Positions shewn.

The Instrument, drawn on Paper stuck to a Rectangular Wooden Board, or engraved on Metal, requires the following Fitments:

1) a Pair of Sights, above, and parallel to, the Top of the Trigon 'hej' (i.e. the Line 'hfj').

2) an Hinged Cursor, whose Tip can be set, *firmly,* over any Point within the Trigon 'hej'.

3) a Plumbline, hanging from the Tip of this Cursor, with a little Pobble-bead sliding stiffly along its length.

In Use, the Tip of the Cursor is set to the Point within 'hej' appropriate to both Latitude and Solar Declination; the Plumbline is drawn over the corresponding Degree of Solar Declination of the smaller Scale 'kl', and the Pobble-bead slid to this exact spot. Thus the large Triangular Scale determines the Point of Suspension for the Plumbline, and the small straight Scale the position of the Pobble-bead along its length. The Sun's Beams are passed (the Instrument being turned in the vertical plane with 'h' towards the Sun) through the Sights, or the Sun being sighted when it casts no Shadow, and the Plumbline swinging free; the Pobble-bead will indicate the exact Time on the Hour-lines within the Table 'pkno'. The Instrument, subject to the selected upper limit of the Scale of Latitudes, is Universal, and may be used anywhere on Earth; either in the Northern or Southern Hemisphæres. At the Equator (under the Cœlestial Æquator) the Tip of the Cursor remains motionless over 'e', the only Seasonal Adjustment being that of the Pobble Bead against the Scale 'kl'. So far as I am aware, and although it appears in the Notebook of Nicholaus Kratzer (Diallist to Henry VIII), the Knowledge of how to construct this Instrument was never popularized in this Country; although Continentals (especially the Germans) constructed it regularly. It is the most useful, simple yet comprehensive, of the many Instruments used to tell the time by Solar Altitude. The *Capuchin Dial* is another, being a simplified Adaption of the above for a single Latitude. The *Ship Dial,* employing (essentially) Lines of Latitude which are Arcs of Circles (because the Latitude Scale goes along a "Mast" hinged at the Centre 'e', and adjustable to a curved Scale of Declinations) is an older variant of the Regiomontanus Dial which, although in practice it is most effective, lacks Geometrical Precision.

ESOTERICS

Die zeit grosz acht/   Künfftigs betracht/
Dasz dich in kürtz   Der tod nicht stürtz.
Der dir zu theil   Gibt ewigs heil.
Oder spricht/ gehe   Ins ewig wehe.

Make much of time/ The future bear in mind/
   That sudden death be not thine overthrow;
Who, when he comes, will bring eternal joy,
   Or say, "Depart to everlasting woe!"

                        (H.W.)

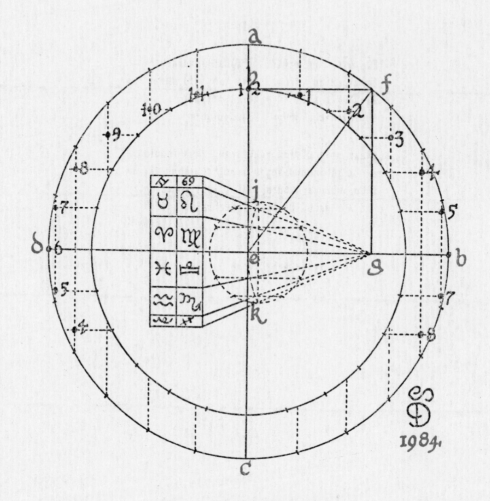

THE ANALEMMATIC DIAL.

52

ESOTERICS

THE ANALEMMATIC DIAL.

So called because its Construction is a direct Application of the *Analemma:* the classical Orthogonal or Orthographic Projection of the Cœlestial Sphære used, although in a much different Application, by the Diallists of the Ancient World. Although usually drawn nowadays from the Works of 18th Century French Mathematicians, the Analemmatic Dial was described a Century earlier in a specific Treatise (*Elliptical or Azimuthal Horologiography,* 1654) by the Englishman *Samuel Foster,* who may well have been its discoverer. Given the usual Prime Circle 'abcd' on Centre 'e'; 'feb' is the Latitude. 'ehfg' forming a Rectangle, 'eh' is the Radius of the Minor Axis Circle. Both Circles are divided into 24 Hours. The Points on the Prime Circle yield the Parallels to 'aec'; those on the Minor Axis Circle the Parallels to 'deb'. Each Hour Point lies where a respective Pair of Parallels cross, and these Points outline an occult Ellipse (which *does not* itself require to be substantiated). The Tangent Scale of Solar Declination lies at 'jek' along 'aec' and is drawn to the Radius 'eg' according to the Principles of the Manæus, and marked with the Signs of the Zodiac. This Scale of the Signs, here expanded past 'jek' for clarity, should in practice be marked on either side of 'jek' and be to that specific Length, with Decantes also. There is no point in marking the uneven and asymmetrical Calendar Months instead of the Proper Zodiac Months as taught, although most modern Writers teach only that practice: it is not so accurate, and is completely unhistorical, lacking also in Æsthetic Simplicity. All except the Zodiac Scale and the Elliptical Sequence of Hour Points should be deleted.

The Dial is Horizontal, and placed with 'a' to the North. Its Gnomon is a Vertical Rod, of adequate but non-specific Length, which is adjustable along the Scale 'jek'. This Gnomon is set to the Sun's Position in the Zodiac on the Day in question, and its Shadow will fall true on the Scale of Hours.

Small Versions should have sliding Tracks under the Dial Plate, with the Gnomon emerging through a Slot along 'jek'; Large Versions have been laid out as Flower Beds, with the Scale as a Path and a Person's own Shadow telling the Hour.

N.B. that Lines drawn from 'e' through the Hour Points form the same Angles, but with 'eb', as those with the Meridian on an ordinary Horizontal Dial.

With a real Understanding of the Principles enunciated in this little Treatise, Analemmatic Dials may be made to all possible Planes, including Declining Planes (in which Instances the Zodiac Scale lies along the "Meridian of the Plane" and not the "Meridian of the Place"; while the 12 o'Clock Point does not lie perpendicularly below the midpoint of the Scale), as required. On

Æquatorial Planes it is the same as an ordinary Æquatorial Dial, with the Zodiac Scale contracted to nothing; whilst on Polar Planes the Scale is expanded to its largest possible Dimensions, and the Ellipse of Hour Points flattened to a Straight Line!

ESOTERICS

Zeit fehrt mit eil/ Hast nich viel weil/
Mach dich bereit/ Zur ewigkeit.

Time goes with haste/ Here's no repose for thee/
Thou must'st make ready/ For eternity.

(H.W.)

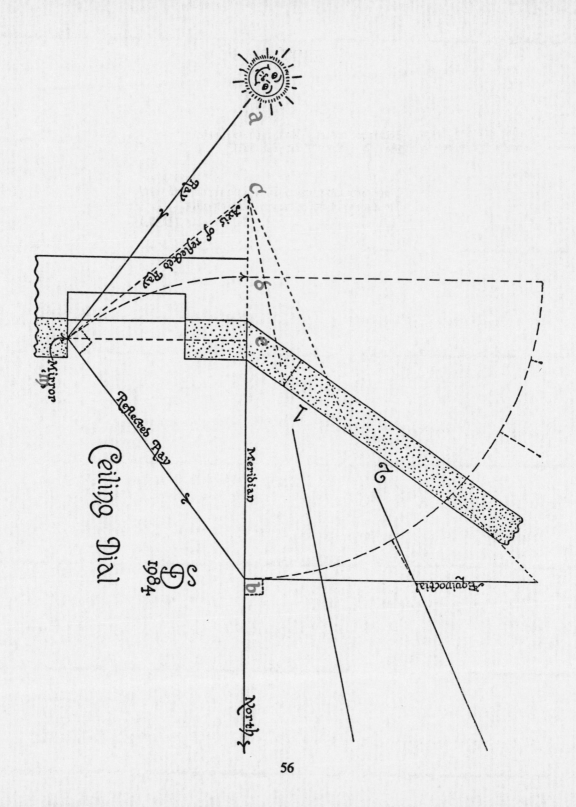

Ceiling Dial

D S
1984

56

ESOTERICS

CEILING DIAL.

BASED on the Principle that a Ray of Light is reflected at the same Angle as it strikes a Mirror, but in the opposite Direction: the Reflective Ceiling Dial is, in essence, no more than the Mirror Image of a very large Horizontal Sundial for the Latitude in question.

'mcb' is the Latitude. At Noon, on the Day of the Æquinox, the Reflected Ray strikes point 'b', establishing, by reference to point 'e', the Meridian and Æquinoctial. The Æquinoctial Hour Points are calculated in terms of the Radius 'mb'; and the Angles which the Hour Lines make with the Æquinoctial are also calculated and set out, so that the Hour Lines themselves can be drawn. It will be noted that, as the Centre 'c' of the Dial lies in the Air outside the Room, less than one half of the complete Hour Circle can be projected; but that the Angle of the Wall of the Room is inconsequential to the Construction of the Dial.

The Mirror should be small and flat, and most precisely placed in a truly horizontal position at the Node of the Projection. The Æquinoctial Line should be produced on the Ceiling with the Hour Lines, and the Hyperbolic Parallels of the Tropics of Cancer and Capricorn are (traditionally) also included as terminators to the Grid of Hour Lines. The reflected Image moves from West to East. The Lines are drawn with a sooted snap-line on the White Ceiling, and substantiated with a Lining Brush after checking for accuracy.

Ceiling Dials were once very common, but most have long since been whitewashed over. The only known original Example is at Milcote Hall Farm, south of Stratford upon Avon, which was uncovered and restored in the late 1940s.

For further Information on the Milcote Ceiling Dial, including a Picture, see Mr. Paul Morgan's Article "A Reflective Sun-dial at Milcote Hall" in the *Transactions of the Birmingham Archæological Society* (1953).

ESOTERICS

Die liebe zeit/ so thewr und werth/
Ohn unterlasz mit eil hin fehrt/
Wer ir recht brauchet wie er sol/
Dem gehts nach diesem leben wol.

Belovèd time/ so worthy and so dear/
    Is hast'ning past, nor halts upon her way/
Who, as he ought to, useth her aright/
    Will live well in the life which lasts for ay.

(H.W.)

Æsthetics

"This eternal blazon"

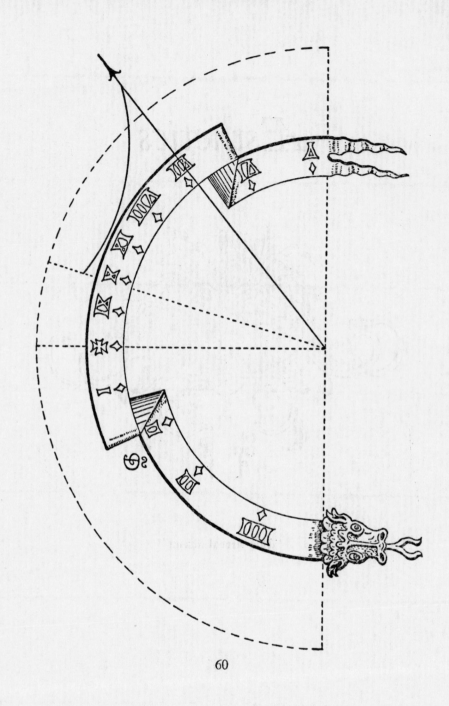

ÆSTHETICS

DIAL PLATES.

HORIZONTAL DIALS are usually made on Circular Metal Plates, with the Geometrical Centre set backwards (about half way between the Centre of the Plate and the Southern Rim), so that the Hour Lines around Noon are well spread out. With Wood or other non-metallic Materials it is better to make them on an exact Square, or perhaps a Lozenge.

VERTICAL DIALS, when not drawn directly upon a Wall, are usually made upon Plates of Stone or Wood (Modern Resin-bonded Waterproof Plywood far excelling any Material of the Past), generally a little taller than an exact Square: the Border, with its Figures and Mottoe, following this general Shape. The older Tradition of using an Horizontal Rectangle, like a Double Square, to accommodate an incised or Painted Semi-circle, did not take root in this Country, although advocated by the earliest Diallists of the 15th/16th Centuries.

The tall Rectangular Stone Dials of the 17th Century frequently have their upper edges surmounted by, or elaborated into, an ornamental Cresting; a Fashion which rapidly died out in the 18th Century.

Declining Dials frequently have their Geometrical Centres offset towards the Side where the Hour Lines are most open, to allow for a more even spread around the Border: this should not be taken to excess.

DIALS DRAWN DIRECTLY ONTO WALLS may, of course, be of any shape. Although the Rectangle predominated in English Dials, Semi-circles are also found. On the Continent of Europe all sorts of Fanciful Shapes were employed. Where a Semi-circle (even sometimes a Rectangle) was chosen, the Border was frequently painted to look like an unrolled Scroll, either with, or without, the incised Hour Lines within.

COLOURING.

ALL DIALS SHOULD BE PAINTED; the modern Prejudice against putting Paint on Stone having no Counterpart in the Past. Only those Dials which are *engraved* on Plates of non-ferrous Metal are unsuitable for this treatment. Sometimes *all* of the Work was carried out in Paint, and this is why many 18th Century Dialstones are now completely blank: it was more usual however, especially during the 17th Century, for the Hour Lines *only* to be actually incised. Some few Diallists incised the Numbers as well. Mottoes, however, were but rarely incised (as that would preclude the possibility of variation); and for that reason are frequently absent to the view.

ÆSTHETICS

The predominant Colour of all Vertical Dials should be White; with the Hour Lines in Black, or perhaps in Red. The Border containing the Figures and Mottoe should be Blue, with Letters and Numbers in Gold Leaf. This is the Concensus of all of the Old Writers on Dialling who mention such things. Other Colours may be employed; but only for effect, or for shading.

THE HOUR LINES may quite properly be terminated by a small Semi-circle or Rectangle, drawn around the Geometrical Centre of the Dial, instead of making a clumsy Black Mess at the Centre itself: if a Rectangle is used some Emblem or Inscription may fittingly be placed inside it. If the Hour Lines are *always* incised, the problem of drawing a straight smooth Line with a Paintbrush is thereby circumvented, the essential Geometric Construction being also preserved from inadvertant alteration during subsequent re-paintings.

HALF-HOUR LINES, when included, should extend in only a little Distance from the Border, and be terminated by a Fleur-de-lis, or by a simple open Trefoil of three dots.

QUARTER HOURS are even shorter, being frequently no more than mere alternating Black and White Portions of a narrow fillet within the Border proper.

THE BORDER should be a broad band all the way around a Vertical Sundial: the Top Portion containing the Mottoe, the Sides and Bottom the Hour Figures.

THE HOUR FIGURES may be either Arabic or Roman. It was quite usual for Arabic Numbers to be used for the Morning Hours, and Roman for those of the Afternoon (rarely vice-versa). Arabic Numbers are to be preferred where the Hour Lines lie closely together. Many early Diallists, and not a few of the later, followed the Mediæval Tradition of using a simple Cross Patty in place of the (always excessively wide) figure 12, or XII, for the "Prick of Noon".

THE MOTTOE should traditionally be pessimistically mournful: displaying either Classical Fatalism or Christian Resignation:

VIGILATE ET ORATE (Watch and Pray)

ASPICE ET ABI (Behold and Begone)

PVLVIS ET VMBRA SVMVS (We are Dust and a Shadow)

ÆSTHETICS

SIC TRANSIT GLORIA [or FABVLA] MVNDI (So passeth the Glory [or Empty Story] of the World)

ΖΩΗ ΑΤΜΗ ΣΚΙΗ (Life is Smoke and a Shadow)

CARPE DIEM ("This is the Day")

LVX VMBRA DEI (Light is the Shadow of God)

MEMENTO MORI ("Remember to Die")

TEMPVS VMBRA (Time is a Shadow)

Etc.

The Genre touches both the Art of the Epitaph, and that of Heraldic Display. A beloved Female Friend has proved, by her own choice, the possibility of a guarded Note of Optimism:

LVCEM SPERO (I hope for Light)

LVX VITÆ (The Light of Life)

TIME TRIETH TRUTH

TEMPVS FVIT EST ET ERIT (Time has been, is, and is to be)

TEMPVS RERVM IMPERATOR (Time ruleth the Affairs of the World)

or

LVX SPEI LVX VITÆ (The Light of Hope is the Light of Life).

Capital Letters are always to be preferred: they should include Æ and Œ where appropriate, and should always employ I and V in place of J and U when the inscription is in Latin. A Central Dot may well be placed between each Word of the Inscription. Excessive Simplicity in the Forms of the Letters is a serious mistake to be studiously avoided. To produce a Classical Roman, or Pure Sansceriph Letter, requires an intense Artistic ability which few possess now, or did possess in the Past: the fluid Capital forms of the 17th Century Monumental

ÆSTHETICS

Mason providing an easier, and much more authentic, Model for Work of this type; in which workmanlike, but distinctly Second-rate, Standards were applicable during the "Golden Age" of the Art in question.

The Colour Blue was notoriously fugitive; being frequently "restored" as Grey or Black.

The Blue used in the 17th and 18th Centuries was made from Blue Bottle Glass, ground to a Powder (called "Smalts") and applied with a Puff to a prepared Surface of sticky White Size. There is no effective Substitute (at least for external use) for genuine Gold Leaf.

Small Portable Dials, drawn with Indian Ink on Paper pasted onto Plates of Plywood (replacing the old Boxwood or Ivory), should be coloured-in with genuine Artist's Watercolours: using only the Earth Pigments, Yellow Ochre, Light Red, Burnt or Raw Sienna, Burnt or Raw Umber, Terra Vert etc., which are fully resistant to the effects of Light. There is firm Manuscript evidence for the use of these Colours by the Diallists of the 16th Century: the Notebook of *Nicholaus Kratzer,* which I have studied (MS.CCC 152 in the Bodleian Library), being full of such examples.

EMBLEMS, suitable for ornamenting Dials, are legion. Most commonly, in this Country, the Sun itself in Gold Leaf with a Face and the Hour Lines emerging from it as Rays, formed the Geometrical Centre of the Dial. In Germany an Human Cranium, accompanied by a suitable Mottoe, was of frequent occurrence; whilst the French employed all sorts of Pretty Devices. The present Writer has used: a Monogram of the Name of the Owner; a Nationalistic Emblem, such as the Thistle; a Coat of Arms; a Flower shedding Petals; and a Winged Cranium (with the Mottoe "I DIE TODAY AND LIVE TOMORROW"), on a little Diptych Dial made for a certain young lady whose name was taken from an old Tombstone. Such Emblems should be drawn with the clear Outlines and crisp Shading of the Sign Painter; avoiding the blurred, muddy, brushmarky effects of the Professional Painter in Oils.

Where this Type of Work is being carried out it is primarily the Effect from a Distance which most needs to be borne in mind. Light and Dark Colours must clearly alternate, except where used as shading: White or Yellow (or Gold) on a Background of Pure Colour is certainly the best combination; things performed in Dark Colours on White or Yellow (avoiding the brightest tones of this Colour when used in mass) need to be much bolder in comparison, as a Light Colour will always spread over a Dark and not vice-versa. Blue should be the Colour of a Cloudless Sky in Summer, without any Trace of Purple; Red should be Scarlet or Vermillion, never Crimson or Magenta. Everything should be primed with Red Oxide, and then brought up with as many Coats of

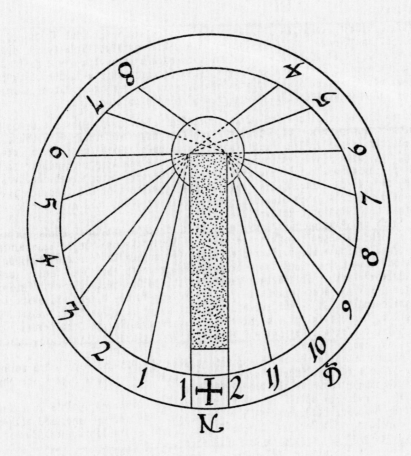

ÆSTHETICS

Gloss as can be managed; sanding down between each Coat. Never use Undercoat externally. Finish all with a Coat of Polyurethane Varnish, from which the Sun will soon extract the Yellow.

Avoid copying, slavishly, the Work of Others.

GNOMONS.

ALTHOUGH, traditionally, the "Gnomon" is taken to be a solid Triangle, consisting of the "Style", actually casting the Shadow; the "Sub-style", directly under this on the Dial's Surface; and an open edge without function; its Essence, and only essential part, is the "Style" (the Straight-edge actually casting the Shadow), and Gnomons which consist of this alone, suitably strutted, are certainly the most effective, as their Shadows are easily read from any angle.

HORIZONTAL DIALS, of course, need something much more substantial, or they will attract the attention of Vandals. Such Gnomons are made very thick, and are securely bolted to the Plate from underneath. They have, in effect *two* "Styles"; i.e. both of the squared-off Edges of the Angled upper Surface: so that the Hour Lines need to be drawn respective to the Edge which actually casts the Shadow at a given Time. Lines drawn behind the Root of the Gnomon (i.e. to the South) will cross each other, as shewn. Only at Noon and at 6 a.m. or p.m. do both Edges function together.

Such an Arrangement may also be applied to Vertical Dials, although there is little purpose in making an high-up Vertical Gnomon so substantially.

Most Old Diallists preferred to make Gnomons for Vertical Dials out of thin Metal Sheet, frequently as a solid Triangle, with the "free Edge" cut to some elaborate set of ornamental Curves as the Maker pleased. Such Gnomons have terrific Wind Resistance, causing them to vibrate, bend, and work loose from their settings. It is best therefore to cut away as much functionless Metal as possible, leaving either a single Strut (either straight or ornamental), or a more elaborate fretted Pattern. Gnomons of Wrought Iron, in various fantastic Shapes, are also to be found.

Where a Nodus is to be included it should take the form of a rounded Notch cut out of the Edge of the Style in such a way as to leave one Sharp Point: the actual caster of the Nodal Shadow.

For small Portable Dials, the best kind of Gnomon possible is a length of strong Thread, held taut with a Strut, or weighted at one end. Diptych Dials, consisting of a Vertical South Plane hinged to an Horizontal Plane, are always fitted with a common Gnomon of this type.

For the elucidation of many of the Points raised in this Chapter, as also for reproductions of Diagrams and Title Pages from old Dialling Books, together with an excellent Bibliography, see *Sundials on Walls* (NMM Monograph 28) by Christopher St. J. H. Daniel (1978).

A DIALLIST'S GLOSSARY

Æquinoctial: The Cœlestial Æquator, drawn as a Circle divided equally into 24 Hours, and projected onto most Dialling Planes as a Tangent Line.

Æquinoctial Radius: The Radius of the Æquinoctial Circle proportionate to the Gnomic Triangle or Rectificatory (q.v.).

Analemmatic: Employing the Principles of the Analemma or Orthographic Projection of the Sphære. See "Orthography".

Dialling: The art of drawing Sundials suitable for any Plane.

Direct Dialling: The Art of drawing Dials on Planes facing due North, South, East, or West. In practice any Dial upon which the Hour Lines may be drawn with no other pre-requisites than those inherent in the Proportions of the Rectificatory (q.v.) may reasonably be styled Direct. This would include the simple Horizontal Dial, with the Polar and Æquinoctial Dials also.

Declining: Turning away from the Direct. Used of Dials drawn upon Planes which do not face due North, South, East, or West; and yet are still Vertical.

Disflect: The same meaning as "Deflect"; but used specifically in Dialling for the deflection of the Sub-style (q.v.) from the Meridian (q.v.) in Declining Dials.

Horizontal: Level with the Horizon. Flat upon the Surface of the Earth, or parallel with it.

Meridian: The Noon, or 12 o'Clock Line, on any Sundial. A Line passing from due South to due North in the Cœlestial Sphære by way of the Zenith.

Nodus: A distinct Point, as opposed to a Straight Edge, casting a Shadow onto the Plane of any Sundial.

Orthography: The Art of drawing anything without Perspective, as if viewed from Infinity. In Dialling the Sphære, so drawn, is composed of Circles, Straight Lines, and Ellipses; the latter, in practice, being avoided as much as possible. Most of the Diagrams in this Book are in, or contain Elements of, the Orthographic Projection. It is the Foundation of all Dialling; as also of most Trigonometrical Formulæ used in Astronomy.

Prime Vertical: A Line in the Sphære connecting the due East point of the Horizon to the Zenith. Also the main Vertical Line in any Diagram.

Radius: The Distance to be taken in the Compasses for the drawing of any Circle.

Rectificatory: A Right Angled Triangle or Set Square whose Subsidiary Angles are those of a specific Latitude and its Complement; containing also the proportionate Radius of the Æquinoctial as a straight line drawn on its Surface from the centre of the Right Angle to form its own right angle with the longest side (Hypotenuse). It was at first fitted with a Plumb Line, for levelling, and used to construct Sundials, empirically, in three dimensions. See Page 12a.

Right Angle: 90°. A dead Square.

Sphære: As so spelt the Geometrical "Great Sphære" of the Heavens is always intended.

Style: That part of a Gnomon (usually a Straight Edge) which casts the Shadow.

Sub-style: A Line on the Dialling Plane marking that part of it which is nearest in Space to the Style. In practice the whole Gnomon is usually fixed along its length.

Stereography: A Projection of the Sphære based on the Half Tangent Scale (where the Tangent of Half a Degree is taken as the Tangent of a Degree; the Half Tangent of 90° being the Tangent of 45°) in which all of the Circles appear as Straight Lines or Perfect Circles. It is fundamental to the Construction of the Astrolabe, but quite impractical in the Construction of Sundials due to the vast Length of some of the Radii required. Some 17th Century Diallists advocated its use, however, and were much citicized by others for so doing.

Trigon: Any Triangular Construction. Specifically the Device described on Pages 12 and 12a.

Trigonometry: "The Measurement of Triangles" fundamental to Dialling, whether carried out, as here, by actual Construction, or by abstract Calculation. The Ratios, all measured in terms of the Radius are;

> To the Angle 'ABC':
> *Radius* is 'CB' (also 'HB' and 'DB')
> *Sine* is 'DE' (also 'FB')
> *Co-sine* is 'DF' (also 'EB')
> *Tangent* is 'AC'
> *Secant* is 'AB'
> *Co-tangent* is 'GH'
> *Co-secant* is 'GB'
> *Chord* is 'CD'

All except the Chord are themselves calculated from 'DI', the *Chord of Twice the Angle*, when expressed in Tables as Figures. Chords were calculated in

68

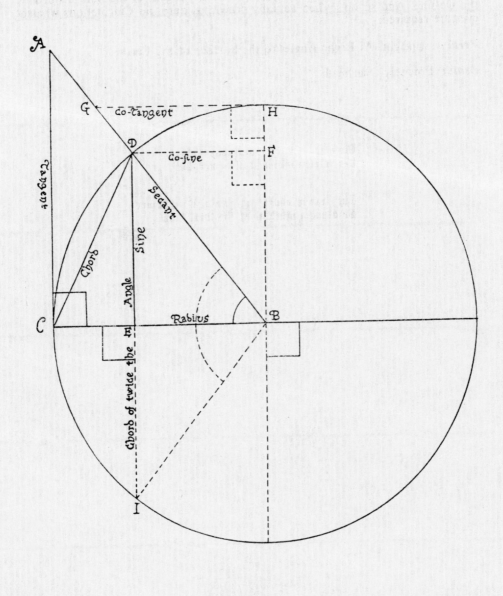

Antiquity. The Sine is half the Chord of Twice the Angle. The Co-sine is half the Chord of twice Complementary Angle. The Tangent is the Sine divided by the Co-sine. And so on. When actually drawn, as here, no Calculations whatsoever are required.

Vertical: Upright. At Right Angles to the Surface of the Earth.

Zenith: Precisely Overhead.

Die zeit ist kurtz/ und ungenwisz/
Der letzten stund ja nicht vergisz.

The time is short/ uncertain is its dower/
Be always mindful of the final hour.

(H.W.)

SELECT BIBLIOGRAPHY

1. *The Notebook* (ined.) of Nicholaus Kratzer (MS CCC 152 in the Bodleian Library). The late Mediæval Dialling Tradition; largely copied from an "old book" in the Library of the Charterhouse of Maurbach (near Vienna), prior to Nicholaus's coming to England. In Latin. See also No. 11.

2. *Horologiographia* (1531) by Sebastian Münster. From similar, but fuller, Sources to those used by Kratzer. A splendid Eposition of the Art by an important Scholar of the Reformation. In Latin, and reprinted (with different Titles) several times. A near contemporary Edition in German is also in existence.

3. *Conformatio Horologiorum Sciotericorum* (1576) by Herman Witekind. Directions for calculating the Requisites for making Sundials on all Planes by Mathematical Calculation (Trigonometry). Not always correct. The actual Hour Lines are drawn by geometry. In Latin.

4. *Horologiographia: The Art of Dialling* (1593) by Thomas Fale. The best and earliest Book on Dialling in the English Language. Essentially a corrected and sensibly expanded Version of No. 3. The Seminal Work.

5. *The Art of Dyalling in Two Parts* (1609) by John Blagrave of Reading. A most entertaining Rag-bag of practical Dialling experiences. Circulated in Manuscript for many Years prior to its publication. Fale refers to the Work in his own Preface. Blagrave's Monument in St. Lawrence's Church, Reading, is still extant.

6. *Horometria: or, The Compleat Diallist* (1650) by Thomas Stirrup. "the whole mystery of the Art". Printed by the Author of 7.

7. *The Art of Dialling* (c. 1669) by William Leybourne. The first Edition of what became (during the Author's long Lifetime) a vast Scholarly Treatise. An excessive Veneration for the awkward Stereographic Projection. Not recommended.

8. *Mechanick Dyalling* (1668) by Joseph Moxon. Sundial Construction for those of "Ordinary Capacity". Shamelessly plagiarized by Charles Leadbetter in the following Century.

9. *Elliptical or Azimuthal Horologiography* (1654) by Samuel Foster. The long-neglected Seminal Work on the Analemmatic Sundial.

Only one of the many "Modern" technical Dialling Manuals is currently still in print:

10. *Sundials: their Theory and Construction* (1973) by Albert E. Waugh. Distinctly the "American Angle". Includes a melancholy Section on the setting up of old Horizontal Sundials bought in Europe for the much lower Latitudes of the United States. Later Editions have been carefully cleared, by the Author, of the several Misprints and Errors of the First.

The following, although not technical Dialling Manuals, are also of Interest:

11. *Sundials at an Oxford College & The Pelican Sundial: Description of the Tables* (1979 & 1980) by Philip Pattenden. Essentially a single Work in two Volumes. A good read and most accessible. Contains an excellent Biography of Nicholaus Kratzer.

12. *Sundials on Walls* by Christopher St. J. H. Daniel (several Editions from 1972 to 1980). A well-illustrated introduction to the Art of the Sundial, with an exhaustive Glossary of the many Technical Terms liable to be encountered. Warmly recommended.

INDEX NOMINUM

FINIS

74